AF871625

Nano Green

T.S. AVINI

Nano Green

Revolutionizing Eco-Solutions with Nanotech

by T.S. AVINI

Table of Contents

Chapter 1: Introduction to Nano Green: Bridging Nanotechnology and Sustainability

As we stand at the dawn of the 21st century, humanity faces an intricate tapestry of challenges, each intertwined with the other in an intricate dance of cause and effect. Climate change, resource depletion, and environmental degradation form the silhouette against which our most urgent battles are waged.

Yet, even as these colossal challenges loom large, innovation flickers like a precious candle in darkness, promising paths to a future that is both sustainable and prosperous. Among these innovative realms, none is more promising, more exhilarating, and more revolutionary than the marriage of nanotechnology and sustainability—the realm we call Nano Green.

Nanotechnology, with its power to manipulate matter at the molecular and atomic levels, emerges as both artist and engineer, crafting solutions with meticulous precision.

Imagine materials a hundred thousand times smaller than the diameter of a human hair, each particle engineered with a purpose—a purpose aimed not just at progression but at the harmonious blend of advancement and conservation. This is the promise of Nano Green, where every innovation is a step towards ecological balance, each breakthrough a nod to Earth's resilience and renewal.

The concept of Nano Green is neither science fiction nor mere academic discourse; it is a burgeoning reality.

The synthesis of these two seemingly disparate arenas—nanotechnology and sustainability—affords an opportunity for unprecedented applications, setting the stage for ecological rejuvenation. But before we embark on exploring these applications, it is crucial to delve into the very essence of the two domains we seek to unite.

Nanotechnology, at its core, is the manipulation of matter on a nanoscale, a dimension that defies the ordinary principles of physics and chemistry. At these minuscule scales, materials exhibit novel properties, opening a Pandora's box of possibilities that scientists and engineers are only beginning to explore.

The color of substances, their strength, their conductivity, and their reactivity all behave distinctly when altered at the nanoscale. This ability to tailor properties at an atomic scale is what gives nanotechnology its edge, transforming traditional scientific constraints into a playground of potential.

On the other side of the spectrum lies sustainability, a pressing mission aimed at meeting the needs of the present without compromising the ability of future generations to meet their own.

Sustainability implores us to harmonize our economic goals with ecological preservation and social responsibility, urging a balance that has historically eluded the trajectory of human progress. It demands innovative solutions to ensure that growth remains equitable, viable, and within planetary boundaries.

As strange as it sounds, In bridging these realms, Nano Green plays a crucial role, offering dynamic solutions to longstanding sustainability challenges. Consider, for instance, the global energy conundrum—a domain where nanotechnology intercedes with transformative potential.

Solar cells, traditionally limited by efficiency constraints, are revolutionized through nanostructures that harness sunlight more effectively, paving the way for energy abundance. Nanocatalysts streamline chemical reactions, reducing energy consumption and elevating efficiency across sectors, while nanomaterials replace scarce resources with sustainable alternatives, enhancing the longevity and recyclability of products.

Beyond energy, the Nano Green revolution propagates through agriculture and water management—domains equally fraught with environmental stakes. Nano-fertilizers and pesticides promise precision agriculture, optimizing resource use while minimizing ecological footprint.

These nanomaterials, engineered for controlled release, ensure that crops receive nutrients in exact increments without overwhelming surrounding ecosystems. In water management, nanotechnology crafts minuscule filters, purifying water with unprecedented efficacy, an innovation that promises access to clean water in even the most resource-strapped regions.

The journey of Nano Green is far from linear, traversing ethical, social, and economic landscapes that necessitate thoughtful navigation. Ethical considerations, particularly concerning the environmental and health implications of nanoparticles, require vigilant examination.

Regulatory frameworks must evolve in lockstep with innovation, ensuring the safety and transparency of Nano Green solutions. Moreover, socioeconomic inclusivity is of paramount importance, as the fruits of such advanced technologies must be accessible across geographic and economic divides to ensure true sustainability.

It's worth noting that As we embark on this exploration, we will encounter tales of ingenuity—where nanotechnology reshapes industries, conserves resources, and rejuvenates communities.

We will delve into success stories and emerging frontiers, drawing a roadmap that navigates the strategies, challenges, and victories of Nano Green initiatives worldwide.

This narrative, however, is not just one of science and technology; it is a narrative entrusting you—the reader—to ponder, engage, and contribute your own imagination and insights to this profound chapter in human history.

The burgeoning field of Nano Green invites participation, whether through innovation, advocacy, or the simple act of informed conversations, acknowledging that every individual holds the capacity to influence the direction of our collective future.

In this book, we aim to journey across the vast landscapes of potential that nanotechnology paints for a sustainable world, unveiling the profound possibilities of Nano Green. The subsequent chapters will peel back the layers of complexity, shedding light on the synthesis of these fields, and highlighting transformative applications that advance both technological frontiers and environmental stewardship.

Together, let us embrace this synthesis of nano-scale precision and eco-conscious vision, illuminating pathways for a future where growth is symbiotic with the planet we call home. Welcome to Nano Green.

Chapter 2: The Dawn of Nanotechnology: A Historical Perspective

In the vast tapestry of technological evolution, there occasionally emerges a thread so fine and revolutionary that it alters the entire weave.

Nanotechnology is such a thread, its emergence heralding a new era of innovation, yet its roots reaching deep into the annals of history. To understand where this remarkable technology is headed, one must first trace where it has been—a journey that spans decades and traverses the boundaries of imagination and scientific rigor.

The concept of manipulating matter at an atomic level didn't burst forth as a singular, spontaneous revelation. Instead, it simmered for decades, slowly collecting the intellectual and experimental energy needed to ignite a true revolution.

One can argue that the dawn of nanotechnology began on a quiet December day in 1959, at the annual meeting of the American Physical Society at Caltech. Here, in a lecture hall teeming with forward-thinking physicists and restless academics, a visionary named Richard Feynman delivered a talk entitled "There's Plenty of Room at the Bottom. ".

At first glance, Feynman, renowned for his ability to elucidate complex concepts with clarity and charm, posed a tantalizing question: what if we could directly manipulate individual atoms? At the heart of his vision was the ability to arrange atoms one by one, creating materials and devices unimaginable with existing technologies.

It was a proposition as audacious as it was revolutionary—a call to expand our ambit from the macroscopic to the microscopic, tapping into potential previously locked away by the limits of conventional science and engineering.

Feynman's talk served as a clarion call to researchers, but the path to actualize his vision was not immediate. It required the development of new tools and the challenging of existing paradigms.

The 1980s marked significant progress in this regard, with the invention of the Scanning Tunneling Microscope (STM) by Gerd Binnig and Heinrich Rohrer at IBM's Zurich Research Laboratory. This remarkable invention, which earned its creators a Nobel Prize in Physics in 1986, allowed scientists, for the first time, to visualize surfaces at an atomic scale. With the STM, manipulating atoms was no longer a theoretical exercise but a thrilling possibility.

Drawing inspiration from the first-time ability to 'see' individual atoms, the field began to burgeon. The 1990s saw the coalescing of various scientific disciplines—physics, chemistry, biology, and materials science—into the unified field we now recognize as nanotechnology.

This interdisciplinary fusion prompted a proliferation of research efforts and impactful discoveries. The manipulation of carbon into various structures, such as buckyballs (discovered in 1985 by Kroto, Smalley, and Curl) and carbon nanotubes (unveiled in 1991 by Sumio Iijima), demonstrated the extraordinary properties that materials can exhibit at nanoscale dimensions.

It's worth noting that This era of discovery was not confined to laboratories. Governments around the globe began to recognize the transformative potential of nanotechnology. In the early 21st century, substantial national investment surged, notably in the United States with the National Nanotechnology Initiative (NNI) launched in 2000.

This coordinated effort by multiple U. S. government agencies signified nanotechnology's transition from speculative science to a field ripe with practical applications. The NNI highlighted areas such as health, environmental sustainability, and industry—domains where the impact of nanoscale innovation promised unprecedented advancements.

The realization of nanotechnology in practice necessitated not just an understanding of atomic placement, but also the properties materials exhibit at this scale. A pivotal discovery was the paradigm shift brought about by quantum mechanics.

At the nanoscale, the classical physics that governs macroscopic behavior gives way to quantum phenomena. These include quantization of energy, quantum tunneling, and superposition—concepts that can endow nanoscale materials with unique and often counterintuitive properties. For example, nanoparticles of gold can appear red or blue depending on their size, a stark contrast to the metallic hue we know.

By the early decades of the 21st century, nanotechnology began to transition from scientific curiosity to commercial reality. The practical implications of harnessing nanotechnology in eco-solutions became evident.

Companies, research institutions, and governments started to focus on the potential for nanotechnologies to address global challenges, from environmental degradation to energy shortages. By fostering the development of more efficient solar cells, water purification systems, and environmental sensors, nanotechnology promised to play a crucial role in steering the world toward sustainable solutions.

This historical narrative illuminates a technology still in its adolescence, yet its potential appears as boundless as the atomic landscapes it explores.

As we stand at the brink of a future perhaps characterized by nanoscale wonders, we do so with a respect for the visionaries and pioneers whose contributions have paved the way. It is their legacy that propels us forward into realms of invention and discovery limited only by the extent of our imagination.

The dawn of nanotechnology is marked by the convergence of ambition and invention, foresight and fortune. It is a triumph of human curiosity—an assembling of knowledge across disciplines to forge tools that better serve humanity and its habitats.

As we proceed further into understanding and harnessing the full capabilities of the nanoscale, the seeds sown in the 20th century promise a harvest of hope and innovation, revolutionizing eco-solutions and reshaping the very fabric of our future on this planet. As this narrative unfolds, one can only anticipate the symphony of innovations awaiting their overture in the grand theatre of the nanoworld.

Chapter 3: Understanding the Basics: What is Nanotechnology?

Nanotechnology is a term that might evoke images of microscopic robots tirelessly working at tasks unseen to the naked eye, or of magical materials transforming our world in unimaginable ways.

While the reality is grounded in science rather than fiction, the potential and the applications of nanotechnology are indeed vast and transformative. To understand this revolution, we need to delve into the very basic question: What is nanotechnology?.

At its core, nanotechnology involves the manipulation and control of matter on a nanometer scale. A nanometer is one-billionth of a meter, or about 80,000 times thinner than a human hair.

When you operate at such a diminutive scale, materials begin to exhibit properties and behaviors that are markedly different from those they possess in their bulk form. This is due to the principles of quantum mechanics which dominate the nanoscale world, allowing elements and compounds to be manipulated in exciting and innovative ways.

As strange as it sounds, To break it down further, consider how materials such as carbon show unique characteristics when configured into nanostructures.

Carbon atoms can be arranged into hollow spheres, tubes, or sheets, resulting in fullerenes, carbon nanotubes, and graphene, respectively. These structures have spurred a wave of research and innovation due to their exceptional strength, electrical conductivity, and thermal properties that are far superior to their bulk counterparts.

The exploration and application of these materials involve a fusion of disciplines, blending physics, chemistry, material science, and engineering into a single forum of boundless possibilities. This is the crux of nanotechnology: the convergence of multiple scientific paths to explore the uncharted territories of ultra-small dimensions.

Yet, while the scale is minute, the implications are anything but small. The unique behaviors observed at the nanoscale open the doors to innovative solutions across various fields including medicine, electronics, energy, and the environment.

Let's take a journey through some key principles that constitute the foundation of nanotechnology. One of the most significant is surface area to volume ratio.

When particles are reduced to the nanoscale, the surface area relative to their volume increases exponentially. This enhanced surface area means that nano-sized particles can react more efficiently, making them perfect catalysts in chemical reactions. This principle is already being harnessed in the development of catalytic converters that reduce emissions from vehicles and in methods that accelerate the breakdown of pollutants in wastewater.

I've seen this happen before: Another fascinating element of nanotechnology is quantum confinement. As particles decrease to the nanoscale, the way electrons move within them changes, leading to modifications in optical, electronic, and magnetic properties.

This is a cornerstone in the burgeoning field of nanoelectronics, enabling the development of transistors and memory storage devices orders of magnitude smaller and more efficient than current models. Flash memory, found in devices like smartphones and cameras, exemplifies how nanotechnology is seamlessly woven into everyday life.

In the realm of medicine, nanoparticles are increasingly playing a starring role. Consider the potential for drug delivery systems where treatments can be precisely directed to disease-affected cells, minimizing side effects and maximizing effectiveness.

Through such targeted therapies, conditions once deemed untreatable can potentially be managed or cured. For example, nanoparticles are being researched as vectors for cancer drugs, designed to target and destroy tumor cells while sparing healthy tissues.

Further illustrating the transformative capacity of nanotech, the environmental applications are alluringly promising.

Nanomaterials can be used to develop efficient methods of water filtration and purification. Nano-scale filters have the potential to remove contaminants, pathogens, and heavy metals from water sources, offering a beacon of hope for areas around the world struggling with access to clean water.

At first glance, Moreover, in the pursuit of renewable energy solutions, nanotechnology is paving new roads.

Solar cells, bolstered by nanotech, can harvest sunlight more efficiently and cost-effectively than traditional methods. This makes solar technology more accessible and a viable contender in the global race for clean energy.

However, with this power comes the responsibility of understanding and addressing the ethical, environmental, and health implications associated with nanotechnology.

As these materials interact with biological systems, their ecological impacts require thorough investigation. Regulatory frameworks need to evolve quickly to keep pace with technological advances and ensure that the environmental benefits outweigh any potential risks.

It is a fascinating time for nanotechnology, which sits comfortably at the intersection of discovery and application.

As researchers continue to explore the complexities and unlock the secrets of the nanoscale world, the implications for ecological solutions remain profound. In 'Nano Green', the potential of nanotechnology is to revolutionize Eco-solutions, offering groundbreaking solutions to some of the most pressing environmental challenges of our time.

Interestingly enough, In conclusion, understanding the basics of nanotechnology is akin to peering through a microscope at an intricate and burgeoning frontier.

It is a domain bristling with potential and laden with promise, ready to redefine what is possible in eco-solutions. As we stand on the precipice of this nano-revolution, the journey into the future promises to be as tiny in scale as it is vast in its ability to affect change—a true testament to the adage that sometimes, great things indeed come in small packages.

Chapter 4: Nanomaterials: The Building Blocks of the Future

In an era where humanity stands at a crossroads, the confluence of innovation and sustainability becomes no longer a choice but a necessity.

At the forefront of this new age are nanomaterials—microscopic titans that hold the promise of transforming the world we live in. From restructuring industries to reshaping lives, these minuscule marvels embody a quantum leap in our quest for greener pastures.

At their core, nanomaterials are particles that exist on the nanoscale, typically between 1 and 100 nanometers.

To put this into perspective, a single sheet of paper is about 100,000 nanometers thick. The unique properties that nanomaterials exhibit—such as increased strength, lighter weight, and superior electrical conductivity—stem from their nanoscale dimensions and the consequential variation in physical and chemical interactions when compared to their bulk counterparts.

The Quantum World Beckons

The advent of nanotechnology propels us into the quantum world, where the rules of classical physics bend to reveal unprecedented opportunities.

This realm is defined by the 'quantum size effect,' where electrons confined within nano-sized structures behave differently, resulting in unique optical, electronic, and magnetic properties. For instance, gold nanoparticles, which typically possess a lustrous yellow hue in bulk form, appear red at the nanoscale—a visual testament to their altered interaction with light.

These transformations are not merely aesthetic.

The ability to fine-tune materials at the atomic level facilitates the development of solar panels with greater efficiency, batteries with extended lifespans, and water purification systems capable of extracting the most minute toxins. This molecular-level mastery is propelling industries towards a sustainable future, providing eco-solutions that are as economically viable as they are environmentally profound.

Catalyst for Innovation

In the quest for sustainability, energy efficiency stands as a pivotal objective. Nanomaterials are stepping into the limelight, acting as catalysts for groundbreaking advancements.

Fuel cells, which convert chemical energy into electricity, have gained significant efficiency boosts through the incorporation of nanostructures. These cells are becoming increasingly viable as alternatives to fossil fuels, particularly in automotives and portable electronics, thus reducing global reliance on non-renewable resources.

Similarly, in the realm of solar energy, nanomaterials are revolutionizing photovoltaic technology.

Traditional solar cells, limited by their energy conversion efficiency and material costs, are giving way to thin-film solar cells and perovskite solar panels. These transformed materials, designed at the nanoscopic level, not only offer improved efficiency but also reduce production costs, making solar power accessible to a wider demographic and extending the grid to remote areas.

Water: Pure and Simple

The world's water crisis looms large, threatening both the livelihoods and the lives of billions globally. Enter nanomaterials, with the potential to revolutionize water purification and desalination processes.

Nanofilters, developed through advanced nanocomposites, innovation, have demonstrated remarkable efficacy in removing contaminants, pathogens, and salts from water sources. By employing carbon-based nanomaterials, such as graphene oxide, these filters achieve high permeability and stability, ensuring long-lasting and efficient filtration systems.

One notable innovation involves the use of nanocellulose, derived from cellulose fibers found in plants, to create sustainable, biodegradable water filters.

These filters require minimal energy input and hold the capacity to make safe drinking water a reality for impoverished communities worldwide, turning the tide on one of the most pressing humanitarian issues of our time.

Health and Medicine: A Frontier Reimagined

Medicine has long been a field defined by its relentless pursuit of precision.

Nanomaterials have redefined the medical landscape, enabling the design of novel delivery systems that target disease with pinpoint accuracy. Nanoparticles engineered for drug delivery can navigate the body's biological pathways with unprecedented specificity, reducing the side effects often associated with conventional treatments.

Cancer therapeutics has particularly benefited from this leap forward.

By exploiting the enhanced permeability and retention (EPR) effect—a phenomenon where nanoparticles tend to accumulate in tumor tissues—researchers have developed nanoparticulate systems that deliver therapeutic agents directly to cancer cells, sparing healthy tissues in the process. This level of precision not only enhances treatment efficacy but also diminishes patient suffering.

I've seen this happen before: Additionally, nanomaterials are integral to the development of diagnostic tools that are faster, more accurate, and less invasive.

Technologies based on nanosensors, which detect biomarkers at incredibly low concentrations, are propelling early detection methods and improving prognoses by allowing for timely intervention.

Redefining Construction

The construction industry, notorious for its environmental footprint, is turning to nanomaterials to pave the way for greener infrastructures.

Nanocomposites, which combine nanoparticles with traditional materials like cement and steel, are resulting in structures that are stronger, lighter, and more durable. By enhancing the thermal properties of building materials through nanotechnology, architects and engineers are crafting buildings that are not only more energy-efficient but also more resilient in the face of a changing climate.

Interestingly enough, Moreover, self-cleaning surfaces, made possible by nanoscale coatings that repel dirt and bacteria, are emerging as key players in the battle against pollution.

These coatings, inspired by the lotus effect—named after the lotus plant's ability to repel water and dirt naturally—are being applied to skyscrapers and public spaces, reducing maintenance costs and curtailing the use of harmful cleaning agents.

Embracing Sustainability

The potential of nanomaterials to amplify sustainable practices is boundless, yet their development is not devoid of challenges.

Concerns about the environmental and health implications of nanoparticles, should they escape into ecosystems, necessitate rigorous assessment and regulation. As such, the advancement of this technology must be accompanied by comprehensive studies to ascertain the role of nanomaterials in our biosphere.

It's worth noting that Collaboration between scientists, policymakers, and industries is crucial to fully harness nanotechnology's promise while ensuring that safety remains paramount.

The dialogue is evolving, focusing on creating frameworks for responsible innovation that acknowledges not only the advances but also the ethical considerations involved.

Conclusion

Nanomaterials are redefining the limits of possibility, acting as veritable building blocks of the future.

As these invisible pioneers continue to bridge the gap between necessity and invention, their influence will merit awe matched only by their measurable impact on creating a sustainable planet. This nanoscale revolution is more than just a scientific enterprise; it is a testament to human ingenuity—our capability to look beyond what is visible, envision what is needed, and build a world that thrives in harmony with nature.

It's worth noting that As we journey further into this nano-age, the tiny yet titanic presence of nanomaterials will echo through industries and societies, urging us toward an interconnected future where innovation and responsibility converge on the threshold of sustainability.

The task ahead is challenging, yet filled with potential—a future constructed not of bricks and mortar but of atoms and molecules, each meticulously engineered for a better tomorrow.

Chapter 5: Green Chemistry: Nanotechnology's Role in Sustainable Practices

In a world where sustainability has transitioned from a mere buzzword to a clarion call for action, the intersection of green chemistry and nanotechnology emerges as a beacon of hope.

Imagine a realm where chemicals are crafted not just with efficiency in mind, but with an eye towards ecological harmony—a world where the materials that fuel our industries, heal our bodies, and empower our electronics are as benign as they are efficient. Welcome to the cutting edge of green chemistry, empowered and expanded by the innovations in nanotechnology.

The Foundation of Green Chemistry

Let's not forget, Green chemistry is anchored on the principle of designing products and processes that minimize the use and generation of hazardous substances. This progressive approach reimagines chemistry not just as a science of molecules, but as an eco-centric discipline that anticipates and prevents pollution and waste at the molecular level.

The 12 Principles of Green Chemistry, established by Paul Anastas and John Warner in the late 20th century, serve as a guide to achieving sustainable chemical manufacturing. Among these principles are the concepts of using renewable feedstocks, reducing chemical derivatives, and designing safer chemicals.

Nanotechnology: A Catalyst for Eco-Innovation

Enter nanotechnology. At its essence, nanotechnology is the manipulation of matter on an atomic, molecular, and supramolecular scale.

It introduces a dynamic shift in how we approach material science, providing tools for precise molecular design. This precision translates into less raw material usage and waste, and often, the substitution of less hazardous substances. It is a perfect partner for green chemistry by enabling the development of innovative solutions that are not only sustainable but also economically viable.

To be honest, #### Nano-Catalysts: Revolutionizing Chemical Reactions

One of the most profound impacts of nanotechnology on green chemistry is the development of nano-catalysts. Catalysts are substances that increase the rate of chemical reactions without themselves undergoing any permanent change.

Nanotechnology enhances what catalysts can do by maximizing their surface area, leading to more efficient reactions and less material waste. Nano-catalysts enable industrial processes to proceed at lower temperatures and pressures, cutting down energy consumption significantly.

For instance, in the production of chemicals such as ammonia, which is a precursor to fertilizers, nano-catalysts can transform traditionally energy-intensive methods into more efficient ones that dramatically reduce greenhouse gas emissions.

This not only reduces the carbon footprint but also the operating costs for chemical plants, supporting a more sustainable industrial model.

Advanced Material Design

Beyond catalysis, nanotechnology empowers the design of advanced materials that make processes more efficient and less harmful to the environment. Consider coatings and packaging.

Traditional methods involve petrochemical-based plastics and coatings that are tough but persist in the environment for generations. Through nanotechnology, biodegradable nano-composites and intelligent packaging materials can be developed, which not only safely degrade but also provide additional functionalities such as barrier properties, UV protection, and even active monitoring for spoilage.

Moreover, nanoparticles, such as titanium dioxide and zinc oxide, have found usage in producing clean and efficient sunlight conversion materials in the form of photovoltaic cells.

These innovations push the boundaries of eco-friendly energy solutions, and as fabrication methods are refined, the potential for scalability in renewable energy becomes more accessible.

Reducing Hazardous Substances

One of the quintessential goals of green chemistry is the reduction of hazardous substances in manufacturing and products.

Nanotechnology offers innovative solutions here by crafting purpose-driven molecules that either replace toxic components or vastly reduce their necessity. For example, in paint formulations, nanoparticles can be engineered to enhance the color and durability, reducing the need for volatile organic compounds (VOCs), notorious for being harmful to both human health and the atmosphere.

Moreover, nanotechnology is at the forefront of developing "smart" materials that respond to environmental stimuli, enabling enterprises to mitigate pollution dynamically.

Materials embedded with environmental sensing nanoparticles can detect and respond to changes in surroundings, adjusting features like permeability and catalytic activity to neutralize pollutants or optimize energy use in real-time.

Sustainability in the Supply Chain

Closely tied with production, the supply chain benefits greatly from nanotech-driven green chemistry.

Reducing material demand through nanoscale efficiencies lessens reliance on resource extraction and associated environmental impacts. This shift results in a healthier balance of resource use and regeneration, critical for the sustainability of ecosystems and economies alike.

Consider packaging, which accounts for a substantial fraction of global plastic production and landfill content.

Nanoscale innovations, such as biodegradable polymers reinforced with nanoparticles, are paving the way for packaging that is both functional and environmentally benign. These initiatives reflect a substantial reduction in waste generation and landfill dependency, showcasing the blending of innovation with accountability.

Challenges and Future Directions

Despite its potential, integrating nanotechnology into green chemistry practices is not without challenges.

The development of safe nano materials demands rigorous testing to understand the long-term ecological and health impacts. It's a delicate balance between innovation and safety, with regulatory measures needed to ensure that nascent technologies don't spawn unintended consequences.

Yet, the promise of nanotechnology as a partner in sustainable chemistry is undeniable.

Future strategies involve more precise regulatory frameworks and international cooperation to ensure safe, widespread adoption. The direction now tilts toward public and private partnerships focused on research, education, and implementation to make sustainable solutions commonplace.

Conclusion

As the horizon of possibilities broadens, green chemistry empowered with nanotechnology holds the promise of a sustainable future.

It invites us to rethink how we craft and interact with the materials around us, urging a shift toward practices that respect and enhance our natural world. Nanotechnology not only supports the fundamental goals of green chemistry but extends them, offering precision and efficiency previously unattainable.

The synergy of nanotechnology and green chemistry is not just a technical evolution; it is a paradigm shift towards a world where products are created for the benefit of both people and the planet.

These advances reflect the determination and ingenuity of humanity and serve as a testament to our capacity to steer industrial practice toward sustainable shores. In this journey, green chemistry, with the aid of nanotechnology, lights a powerful path forward, guiding industries to a future where economic and environmental goals are not mutually exclusive, but tightly interwoven in the fabric of innovation.

Chapter 6: Energy Revolution: Nanotech in Renewable Energy Solutions

As the global community stands at the precipice of an energy revolution, the advent of nanotechnology offers an unparalleled opportunity to redefine our approach to renewable energy sources. The movement toward sustainable and clean energy is not just an idealistic vision; it is a survival imperative.

With climate change continues to threaten ecosystems and economies, the importance of efficient and innovative energy solutions becomes paramount. Enter nanotechnology, a field of science and engineering that has begun to unlock the potential for enhancing the efficiency and feasibility of renewable energy sources.

Nanotechnology deals with structures and devices at the atomic and molecular scale.

By operating on such a minute level, we can manipulate the very building blocks of materials to achieve remarkable enhancements in energy production, storage, and efficiency. This precision opens the doors to groundbreaking applications in solar power, wind energy, battery storage, and even bioenergy.

Let us first explore solar power, an arena where nanotechnology is truly making waves. Traditional silicon-based solar panels have been the mainstay for decades, yet they face limitations in efficiency and production costs. Nanotechnology offers a leap forward by utilizing materials such as perovskite and quantum dots.

Perovskite solar cells, for example, employ a class of materials known for their excellent photovoltaic properties. With the inclusion of nanoscale engineering, these cells can achieve higher efficiency rates and can be produced at a lower cost. The adaptability in design also allows for the creation of flexible and transparent panels, potentially revolutionizing urban landscapes by incorporating solar harvesting capabilities into windows and building facades.

Quantum dots, tiny semiconductor particles only a few nanometers in size, present another promising avenue. By manipulating their size, we can fine-tune their ability to absorb and emit light.

This capability translates into solar panels that can harness a broader spectrum of sunlight, thereby maximizing energy capture. Furthermore, quantum dots can be integrated into thin films, providing a lightweight and flexible alternative to bulky panels without compromising on performance.

Beyond solar energy, nanotechnology is also redefining how we capture and store wind energy. The traditional wind turbine, though effective, encounters issues related to wear and tear and efficiency losses due to mechanical resistance.

Here, nanotechnology steps in with solutions such as nanostructured coatings for turbine blades, which reduce resistance and enhance durability. These coatings, engineered at the molecular level, exhibit superhydrophobic and self-cleaning properties that combat dirt accumulation and weathering, leading to longer-lasting and more efficient turbines.

The role of nanotechnology in energy storage is just as pivotal. With an ever-increasing reliance on intermittent renewable sources, efficient storage systems are crucial for consistent energy supply. Nanotechnology is at the forefront of innovations in battery technology, particularly with lithium-ion and lithium-sulfur batteries.

The inclusion of nanostructured electrodes and gel electrolytes addresses issues of capacity, charge speed, and longevity. By increasing the surface area and conductivity of battery materials, nanotech facilitates faster ion exchange, resulting in quicker charging times and improved energy density. This advancement is critical for both consumer electronics and larger grid storage solutions.

Nanotechnology also extends its reach to the realm of bioenergy. Here, the focus is on optimizing the breakdown and conversion of biological materials into energy.

Nano-catalysts, engineered to enhance biochemical reactions, offer improved efficiency in biofuel production. These catalysts enable faster conversion of biomass into fuels while minimizing environmental impact. Additionally, nanostructured enzymes can be tailored to break down lignocellulosic materials, previously considered waste, into valuable energy resources.

One cannot overlook the impact of nanotechnology in geothermal and tidal energy sectors, though perhaps less publicized.

In geothermal energy, nanoscale sensors are used to provide precise data monitoring, optimizing the efficiency and safety of heat extraction processes. Tidal and wave energy systems also benefit from nanocoatings that enhance the resilience and longevity of underwater machinery.

Let's not forget, Despite the potential, the integration of nanotechnology in renewable energy solutions is not without challenges.

Concerns about the environmental impact of manufacturing and disposal of nanomaterials need addressing. As we push forward, it is imperative to establish guidelines and standards for the development and deployment of nanotechnology in energy applications, ensuring that the solutions themselves do not become sources of environmental harm.

The road ahead is promising. Nanotechnology holds the key to not merely improving, but transforming how we generate, store, and consume energy.

The fusion of nanotechnology and renewable energy represents a symbiosis of science and sustainability, where innovation meets responsibility. As researchers and engineers continue to push the boundaries of what is possible, we stand on the brink of an energy revolution that could power the world sustainably for generations to come, ensuring a thriving planet for all its inhabitants.

Chapter 7: Water Purification: Nanoscale Approaches to Clean Water

---

Chapter 7: Water Purification: Nanoscale Approaches to Clean Water

Believe it or not, The abundance of water, which is paradoxically both a life-giving force and a destructive element, is a defining feature of our planet. Yet, despite the ubiquity of the blue on our global map, clean water access remains a pressing challenge, especially as populations soar and climates change.

Enter nanotechnology—a field of innovation poised to redefine the parameters of environmental sustainability. In this chapter, we delve into the transformative potential of nanoscale approaches to water purification, an arena where the minuscule reigns mighty.

At the intersection of chemistry, biology, and technology, nanotechnology offers novel methodologies that transcend the limitations of traditional water purification processes.

As we navigate the complexities of water contamination—ranging from microbial pathogens to dangerous chemical pollutants—nanotechnology stands out not just as a hero in filtration but as an ally in creating sustainable and efficient systems.

Unleashing the Potential of Nanomaterials

Nanomaterials, defined by their ultra-small size and large surface area, possess unique properties that traditional materials do not. This includes enhanced reactivity and conductivity, making them ideal for targeting and neutralizing water contaminants with precision.

One such innovation is the advent of nanosilver particles. Known for their antibacterial properties, these nanoparticles can disrupt bacterial membranes, rendering them inactive and thus, purifying water. Their efficiency, coupled with their minimal impact on human health and the environment, positions nanosilver as a cornerstone of nanotechnology-driven water purification.

Similarly, nanotubes and nanofibers are gaining traction for their ability to filter pollutants at the molecular level. Carbon nanotubes, with their robust structural integrity and incredible absorptive capabilities, can effectively remove heavy metals and organic compounds from water sources.

This is achieved through adsorptive interactions, where contaminants adhere to the extensive surface area of the nanotubes, effectively trapping them from drinking supplies. The beauty of these carbon-based wonders lies not just in their efficacy but also in their reusability, making them a cost-effective solution for large-scale water treatment.

Graphene: The Wonder Material

In real life, Of all the nanomaterials explored in water purification, graphene is perhaps the most talked-about.

This one-atom-thick allotrope of carbon is celebrated for its unparalleled properties, including exceptional strength, conductivity, and impermeability. Researchers are now harnessing the potential of graphene-based membranes, which are capable of facilitating water desalination and filtration with unprecedented efficiency.

Graphene oxide membranes, in particular, have shown promise in desalination efforts.

These membranes exploit the two-dimensional nature of graphene, allowing water molecules to pass through while halting salt ions and other pollutants. This selective permeability not only purifies but also addresses the energy-intensive bottleneck of traditional desalination methods, presenting a more sustainable approach to cultivating drinkable water from saline environments.

Catalysts of Change: Photocatalysis

Many people don't realize this, but In the realm of nanotechnology-driven purification methods, photocatalysis stands out. This process capitalizes on nanoparticles' ability to act as catalysts under specific light conditions, breaking down complex pollutants into benign substances.

Titanium dioxide nanoparticles, for example, activate under ultraviolet light to spur chemical reactions that neutralize organic pollutants and pathogens. This self-cleaning property, powered by sunlight, makes photocatalysis particularly attractive for remote or off-grid applications, ensuring clean water without the need for extensive infrastructure.

The Challenge of Scale and Implementation

Despite the enormous promise these technologies hold, transitioning from laboratory success to real-world implementation poses significant challenges.

One of the primary hurdles is scalability—ensuring that these nanoscale solutions can be produced and deployed cost-effectively on a global scale. The manufacturing processes, while innovative, must be streamlined to avoid environmental drawbacks themselves, adhering to principles of green chemistry.

To be honest, Furthermore, the environmental and health implications of introducing nanomaterials into ecosystems are still under exploration.

Careful life cycle assessments and rigorous testing are essential to ensure that these solutions do not inadvertently introduce new forms of pollution or toxicity.

Regulatory and Social Dimensions

As with any groundbreaking technology, regulatory frameworks must evolve to encompass the nuances and possibilities presented by nanotechnology.

Clear guidelines are required to monitor the production, disposal, and potential environmental impact of nanomaterials used in water purification. Collaboration between scientists, policymakers, and industry leaders is crucial in crafting regulations that safeguard public health while fostering innovation.

Public perception and acceptance also play pivotal roles.

Education and transparency are vital in demystifying the science behind nanotechnology, instilling confidence that these advanced methods are safe and beneficial. Integrating community engagement and participatory approaches can enhance the social acceptance of nanotechnology in water purification systems.

A Vision for the Future

In envisioning a future where clean water is a universal reality, nanotechnology emerges as an essential ally.

The nanoscale approaches we've explored in this chapter offer a glimpse into how science can address some of the most pressing ecological challenges of our time. By harnessing the power of the very small, we can achieve monumental impacts.

Interestingly enough, The journey towards wide-scale implementation and acceptance of nanotechnology in water purification is ongoing, and it demands a multidisciplinary effort.

However, with continued research and collaborative innovation, we're poised to transform the landscape of water accessibility and sustainability. As we forge ahead, embracing these technologies, we open doors to lives unhindered by water scarcity, communities empowered with access to safe drinking water, and ecosystems rejuvenated by clean and abundant water sources.

In a world where the value of water is as great as its necessity, our pursuit of innovative solutions like nanotechnology embodies the essence of progress and hope.

Herein lies the promise of a future redefined not just by the shadows of our challenges but by the brilliance of our solutions. The nanoscale is not just a dimension—it is a profound opportunity waiting to unfold.

Chapter 8: Agriculture and Food Security: Nano Innovations on the Farm

As the global population burgeons towards the daunting horizon of ten billion souls, the urgency of revolutionizing agriculture to ensure food security has never been more pronounced. Traditional farming methods, proven over centuries, are increasingly inadequate in the face of climate change, resource scarcity, and escalating agricultural demands.

Enter the world of nanotechnology, a burgeoning field poised to redefine how we cultivate, protect, and produce food. Nano innovations offer astonishing possibilities for enhancing agricultural productivity while minimizing environmental impact, heralding a new era of sustainable farming practices.

At the heart of these nano innovations is the ability to manipulate matter on an atomic or molecular scale, allowing for unprecedented control over materials and processes.

This precise level of manipulation confers significant advantages to various agricultural applications, from crop production to pest management and soil health enhancement.

Many people don't realize this, but One of the most promising uses of nanotechnology in agriculture is in the realm of nano-sensors and smart delivery systems. Farmers have long relied on their knowledge and experience to gauge crop health and nutrient needs, often supplemented by time-consuming and costly testing procedures.

However, nano-sensors embedded in the soil or on plants can provide real-time monitoring of vital parameters such as moisture levels, nutrient content, and even the presence of pests or diseases. This data can be transmitted wirelessly to farmers, empowering them to make informed decisions swiftly. These smart systems not only optimize resource use, reducing water and fertilizer wastage but also minimize the environmental impacts associated with over-fertilization and runoff.

In addition to monitoring, precision farming also benefits from targeted delivery systems enabled by nanotechnology. Traditional methods of pesticide and fertilizer application can be inefficient, with a significant portion of these substances failing to reach their intended targets.

Nano-encapsulation of these chemicals allows for their slow and steady release at the desired site, minimizing losses and enhancing effectiveness. Nanoformulations can be designed to respond to specific triggers—such as changes in pH or temperature—ensuring that active ingredients are released only when needed. This results in less chemical input, reduced environmental impact, and, crucially, lower costs for farmers without compromising yield.

But the potential of nanotechnology extends beyond existing practices. Consider, for example, the development of nano-coatings for seeds.

These coatings can protect seeds from pathogens and harsh environmental conditions, increasing germination rates and seedling survival even in less-than-ideal soils. Furthermore, nano-coatings can be imbued with growth-promoting factors, offering young plants a robust start and contributing to more uniform and vigorous crop development.

In real life, Another groundbreaking application lies in plant protection and pest control. Conventional pesticides can have deleterious effects on non-target organisms and ecosystems.

Nano-formulated biopesticides provide a solution by enhancing selectivity and efficiency. By encapsulating active agents within nanoparticles, we can engineer systems that target specific pests without affecting beneficial insects. Some nano-formulations even possess the ability to identify pest presence and respond accordingly, leading to more intelligent and environmentally benign pest management strategies.

Innovations in material science are also vital for developing more efficient agricultural machinery and infrastructure.

Nanostructured materials offer superior durability and efficiency, promising lighter, more robust machinery that consumes less fuel. Similarly, advances in nano-composite materials could lead to more sustainable and efficient greenhouses and storage facilities, reducing energy use while maintaining optimal growing conditions.

Furthermore, the potential of nanotechnology in enhancing soil health cannot be overlooked. The indiscriminate application of chemicals has depleted our soils, reducing their fertility and contributing to environmental degradation.

Nano-clays and mineral formulations promise to rejuvenate these soils by improving nutrient retention and water holding capacity. This not only regenerates the soil's inherent fertility but also enhances its resilience against erosion and nutrient leaching.

It's worth noting that While the promise of nanotechnology in agriculture is undeniably captivating, it is crucial to approach its implementation thoughtfully.

The regulatory frameworks surrounding nanotechnology are still evolving, and there are legitimate concerns regarding the potential impacts of nanoparticles on human health and the environment. Rigorous testing and transparent oversight will be essential to ensuring that nano-innovations enhance our agricultural practices responsibly and sustainably.

Public acceptance is equally important, and here lies both a challenge and an opportunity. Misinformation and fear can hinder the adoption of novel technologies.

Clear communication, education, and engagement with communities are vital to building confidence in nano-scale innovations. Highlighting success stories, demonstrating economic and environmental benefits, and involving stakeholders in the development process can foster trust and enthusiasm.

In conclusion, nanotechnology offers a vibrant and transformative approach to addressing the pressing challenges of global agriculture and food security.

By harnessing the innovative potential of nano-scale solutions, we unlock opportunities for significantly enhancing efficiency, productivity, and sustainability on the farm. As we stand on the cusp of this technological revolution, the future of farming promises not only to feed a hungry world but to do so in harmony with the environment, ensuring that our fields are bountiful today and fertile for generations to come.

Chapter 9: Environmental Remediation: Nano Techniques for a Cleaner Planet

The canvas of our world is painted with a spectrum of wondrous environments—from the azure seas and lush landscapes to the rich minerals beneath our feet. However, the same industrial ambitions that have propelled humanity to remarkable advancements have also cast long shadows across these natural wonders.

Pollution, in its many forms, marbles the surface of our planet, prompting a clarion call for innovative solutions. This is precisely where nanotechnology emerges as a beacon of hope, offering unprecedented promises of environmental remediation—a cleaner planet through the artful manipulation of matter at the nanoscale.

The beauty of nanotechnology lies in its scale—billionths of a meter—where the laws of physics play by different rules, allowing materials to exhibit unique properties. This realm provides an unconventional toolkit for addressing environmental challenges with efficiency reminiscent of nature itself.

Starting with one of the most ubiquitous resources, water, nanotechnology has spearheaded advancements in purification and decontamination. Central to these breakthroughs are nanoparticles coated with functional compounds, designed to target and neutralize pollutants with pinpoint precision.

Consider the pervasive problem of heavy metals such as lead and mercury, tainting water sources with their toxic legacy. Traditional methods falter under the burden of cost, effectiveness, and secondary waste generation.

Enter nanoscale solutions: engineered nanoparticles, such as zero-valent iron (ZVI) and titanium dioxide, act as reactive agents. ZVI reduces metal ions, precipitating them as harmless solids, while titanium dioxide's photocatalytic properties harness light to degrade organic contaminants. These nanomaterials operate with surgical efficacy, attacking pollutants at a molecular level, leaving clear water in their wake.

Beyond purification, nanotechnology extends its influence to air—often the overlooked frontier in pollutant management. Airborne particulates, gases, and volatile organic compounds (VOCs) contribute to urban smog and respiratory diseases.

Here, nanoscale catalysts embedded in air filters expedite the decomposition of hazardous gases. Furthermore, nanostructured photocatalytic coatings transform building surfaces, actively breaking down NOx emissions and organic compounds, turning urban spaces into living air purifiers.

Soil, the earth's nurturing carpet, too bears the weight of pollution through industrial spills, agricultural runoff, and improper waste disposal. Remediation techniques integrating nanotechnology bring a new strength to bioremediation and phytoremediation strategies.

For instance, flag-bearing nanoparticles can be used to deliver nutrient and stimulant packages to soil-dwelling microbes, boosting their ability to degrade hydrocarbons. Similarly, nanoparticles can fortify the absorption capacity of plants, enabling roots to extract and stabilize contaminants more efficiently, accelerating the restoration of barren landscapes to healthy ecosystems.

In real life, At the heart of these revolutionary applications is the principle of specificity. Nanoparticles can be tailored to recognize and bind specific contaminants, a concept borrowed from biological processes like antigen-antibody interactions.

For instance, nanostructures have been designed to mimic 'smart' biomolecules that recognize and isolate oil molecules, separating them from water in oil spill scenarios. This specificity minimizes the collateral damage traditionally associated with remediation, preserving the delicate balance within ecosystems.

Remarkably, the hallmark of nanotechnology in environmental applications is its commitment to sustainability. Traditional remediation processes often rely heavily on energy, introducing secondary waste streams in the quest to cleanse environments.

Conversely, nanotech solutions are designed to be not only efficient but inherently green. Self-assembling materials, biodegradability, and the conversion of sunlight into usable energy typify the nano-assisted environmental solutions being developed. The sustainability ethos extends beyond immediate application, securing the integrity of ecosystems for generations yet unborn.

However, with great power comes great responsibility. The field of nanotechnology is not without its challenges and considerations.

The introduction of nanoparticles into ecosystems prompts a meticulous evaluation of their long-term environmental impact. The intricate balance sought by designers and researchers seeks to mitigate unintended consequences, emphasizing comprehensive risk assessments and the development of safe, inert nanoparticles. Transparent regulatory frameworks must steer the course, establishing guidelines to harmonize innovation with ecological stewardship.

Believe it or not, The promise of nanotechnology in environmental remediation paints a future not only where damage is mitigated but entirely reversed.

This future envisions sewage treatment plants retrofitted with nanofiltration systems that return nutrients back to aquatic ecosystems, mining operations where tailings are safely sequestered and recovered, and oil spills rendered benign within days rather than years. It imagines a world where the footprints of human enterprise are lighter and symbiotic rather than burdensome and exploitative.

Indeed, the revolution in environmental remediation powered by nanotechnology speaks to a profound truth: the solutions to our greatest challenges often lie within the grasp of our tiniest innovations. As we harness this potential, we are poised not only to clean our planet but to redefine our relationship with the environment.

In crafting a cleaner, more sustainable future, nanotechnology does not merely restore—it revolutionizes the very fabric of our ecological tapestry, interweaving human ingenuity with the resilience of nature. Thus, as we stand on the brink of this nanotechnological era, our commitment to a cleaner planet is bolstered by our ability to envision, design, and implement solutions as boundless and nuanced as the nature we strive to protect.

Chapter 10: Air Quality Enhancement: Nanotechnology in Pollution Control

Imagine stepping into a world where the air you breathe is as pristine as a mountain breeze, regardless of whether you are in bustling urban centers or pastoral countryside settings.

It's an appealing dream—one that the evolution of nanotechnology is beginning to make a reality. In an era when air pollution ranks as one of the leading environmental health risks globally, the microscopic wonders of nanotechnology offer profound potential to redefine air quality management—ushering in a new standard for the very air that sustains our lives.

Air pollution is a complex soup of gases, particulates, and biological molecules, all of which compromise the health of ecosystems and humans alike. Carbon dioxide and methane from industrial processes, fine particulate matter from combustion, and volatile organic compounds from numerous sources each play a part in this orchestration of harmful elements.

Traditional mitigation strategies have often been reactive, managing symptoms rather than curing the disease. Enter nanotechnology, with its promise of economically viable, efficient, and proactive solutions to pollution control.

Believe it or not, The potential of nanotechnology in air quality enhancement lies primarily in its materials—nanomaterials, more precisely.

These materials, with at least one dimension in the nanometer range (a millionth of a millimeter), have extraordinary properties that can be harnessed for environmental purposes. Let's delve into several areas where these nano-scale knights are waging war against pollution.

One of the most promising applications is in the development of advanced filtration systems. Traditional air filters, while effective to a certain extent, struggle with the nanoscale particles that can evade these defenses and cause severe health problems.

Nanofibers, due to their high surface area-to-volume ratio, serve as highly efficient sieves that capture these ultrafine particles. Electrospun nanofiber filters, for example, have shown remarkable efficacy in removing particulate matter down to the nanoscale, offering a layer of protection traditional filters cannot match.

In addition to particulate removal, nanostructured catalysts are making headlines for their potential in reducing harmful gaseous emissions. Catalysts accelerate chemical reactions without being consumed, transforming harmful gases into less-harmful substances.

Nanomaterials, such as titanium dioxide (TiO2) nanoparticles, can decompose pollutants like nitrogen oxides and volatile organic compounds when exposed to UV light—a process that can be integrated into building materials, road surfaces, and automotive applications. This has profound implications, especially for urban areas plagued by smog and exhaust pollution.

As strange as it sounds, Photocatalysis, a process that uses light to activate a substance that can cause chemical changes, represents another genius nanotechnological feat in air purification.

Photocatalytic coatings, which can be applied to surfaces such as windows, walls, and roofs, use sunlight to trigger chemical reactions that break down airborne pollutants. Imagine a building that not only consumes zero energy but also purifies the air around it through its very skin—a true marvel of sustainable urban design brought to life by nanotechnology.

Another innovative approach is the development of active air purifiers based on nanomaterials that can adsorb gas molecules.

These devices incorporate carbon nanotubes and graphene, materials known for their exceptional surface area and reactivity, to trap and remove harmful gases from the environment efficiently. These active systems could one day become as ubiquitous as air conditioning units, improving indoor air quality in homes, offices, and public spaces alike.

Nanotechnology not only features in the passive systems of pollution control but is also increasingly incorporated into active monitoring of air quality.

Nanotechnology-based sensors are enabling real-time monitoring of air pollutants at levels of precision and accuracy that were previously unimaginable. These sensors, often small enough to fit inside mobile devices, provide granular data that can inform both individuals and policy makers and facilitate a more responsive and targeted approach to pollution control.

In practice, the integration of these technologies presents a challenge but also an opportunity for interdisciplinary collaboration between nanotechnology developers, government bodies, and environmental organizations.

Regulatory frameworks are crucial to ensuring that these technologies are deployed safely, and that their introduction into the consumer market is guided by sound scientific principles and ethical responsibility.

Incorporating nanotechnology into air quality solutions must occur hand-in-hand with public education and community engagement.

People must understand how these innovations work and what they bring, not only for individual well-being but for collective environmental health. This includes addressing the concerns that come with the development and proliferation of nanotechnology, such as the potential toxicity of nanoparticles—a legitimate concern that must be rigorously researched and managed.

The evolution of nanotechnology represents a remarkable fusion of science and practicality—a harmonious blend of the microscopic and the monumental—with the immense potential to redefine how we tackle one of humanity's most pressing challenges.

By entrusting the oversight of our air's purity to the incredible properties of nanomaterials, we articulate a vision of a future where our atmosphere is shielded from pollution, our cities are free from smog, and the air we breathe is fresh and clean.

At first glance, This chapter illustrates that the solutions to air pollution do not involve looking skyward for answers, but rather, peering into the infinitesimally small, where the brilliance of nanotechnology is unfurling a new era in environmental stewardship.

As we embrace these innovations, let us do so with the awareness that we are not only adopting novel technologies but also stepping toward a future where the art of breathing clean air is a reality for all.

Chapter 11: Nanotechnology in Waste Management: Toward Zero Waste

In an age where environmental sustainability has emerged as a pivotal concern, the horizon of waste management is being dramatically transformed by the minuscule but mighty power of nanotechnology.

The traditional paradigms of waste reduction, recycling, and disposal are increasingly being enhanced by nanotech innovations, steering us ever closer to the ambitious goal of achieving zero waste.

A Shift in Paradigm

It's worth noting that For decades, the approach to waste management has largely been linear—produce, consume, and dispose. This has led to overflowing landfills and ocean debris, with devastating consequences for ecosystems.

The concept of zero waste presents an alternative model: a circular economy where resources are reused, repaired, and recycled indefinitely. Here, nanotechnology is poised not just as a complementary tool, but as a transformative force reshaping the way we understand and manage waste.

Understanding the Nanoscale

To fully appreciate the impact of nanotechnology in waste management, one must first understand the scale involved. Nanoparticles, which measure less than 100 nanometers, possess unique properties that differ from their larger counterparts.

These properties include increased reactivity, strength, and conductivity due to the higher surface area to volume ratio. As a result, nanomaterials can interact with waste in more efficient, effective ways than traditional methods allow.

In real life, ## Nanofiltration: The Forefront of Wastewater Treatment

One area where nanotechnology is already making significant strides is in the treatment of wastewater.

Traditional wastewater treatment processes often struggle with the removal of microscopic contaminants, such as heavy metals and organic compounds. Enter nanofiltration—a process that employs membranes embedded with nanoparticles to filter out these minute pollutants with unprecedented precision.

Nanofiltration not only improves the quality of the treated water but also enhances the efficiency of recycling efforts.

With cleaner effluent, water can be reused in industrial processes, agricultural irrigation, or even potentially as potable water, drastically reducing the need for fresh water resources and thereby closing the loop in water management.

In real life, ## Nanocatalysts: Revolutionizing Chemical Reactions

Waste management is not merely about handling physical refuse; it also encompasses the numerous chemical reactions required to treat and recycle waste.

Nanocatalysts, with their exceptional surface reactivity, are revolutionizing these chemical processes. By accelerating and optimizing reactions at the atomic level, these catalysts facilitate the breakdown of complex plastic polymers or enhance the conversion of organic waste into biogas.

For instance, nanoparticle-based catalysts can degrade polyethylene, a common plastic that typically resists natural weathering and decomposition.

By accelerating such processes, nanocatalysts can help reduce the plastic waste burden that chokes our planet. Similarly, nanocatalysts can enhance the efficiency of biogas production from organic waste, contributing to renewable energy supplies while reducing landfill volumes.

Nano-enabled Sensors: Monitoring Waste Streams

Another promising innovation lies in the development of nano-enabled sensors, which offer real-time monitoring of waste streams.

These sensors can detect and analyze contaminants at ultra-low concentrations, providing immediate feedback on the chemical composition of waste materials. For waste management facilities, this technology enables more accurate sorting and processing, ensuring materials are directed toward the most appropriate recycling or treatment pathway.

Moreover, these sensors can be integrated into waste containers, enabling smart waste management solutions that optimize collection routes, predict overflow, and minimize environmental impact.

The data collected can inform policy decisions and promote community engagement in waste reduction efforts.

Addressing Concerns and Challenges

While the potential of nanotechnology in waste management is vast, it is imperative to navigate its challenges responsibly.

The environmental impact of nanoparticles themselves is an area of active research, as their long-term ecological effects remain incompletely understood. It is crucial to ensure that the benefits of deploying nanotechnology do not come with unintended consequences, such as nanoparticle contamination.

Additionally, the economic feasibility of incorporating nanotechnology into existing waste management infrastructure poses a challenge.

The initial investment can be substantial, and as with many emerging technologies, widespread adoption is contingent upon cost reductions and demonstrable improvements over current systems.

From Innovation to Implementation

The successful integration of nanotechnology into waste management requires collaboration across scientific disciplines, industry, and government policy.

Research and development must be coupled with rigorous safety assessments and robust regulatory frameworks to ensure public and environmental health. Incentives for green innovation, such as grants and tax breaks, could stimulate investment in nanotech solutions.

Furthermore, public awareness and education programs are essential to foster understanding and acceptance of new technologies.

As citizens become stakeholders in the movement toward zero waste, they become empowered to make informed choices that align with sustainable practices.

As strange as it sounds, ## Conclusion: A Nano-Powered Path to Sustainability

The road to zero waste is not without its hurdles, but the promise of nanotechnology paints a hopeful picture for the future of waste management.

By harnessing the power of the very small, we stand on the brink of monumental changes that could fundamentally alter how we interact with waste—transforming it from an environmental burden into a resource for innovation.

As we continue to develop and adopt these groundbreaking technologies, we move closer to a world where zero waste is not just an aspirational goal, but a tangible reality.

In conserving our planet's resources, ensuring clean water, and reducing greenhouse gas emissions, nanotechnology is revolutionizing eco-solutions, offering a pioneering path toward sustainable living. With each nano-sized step forward, we shape a future that values both technological advancement and ecological harmony.

Chapter 12: The Role of Nanotech in Climate Change Mitigation

In an era where the environment sits precariously on the brink of irreversible change, the emergence of nanotechnology presents a beacon of hope. Its potential to revolutionize climate change mitigation is not just promising—it is transformative.

Picture a world where carbon emissions are not merely reduced but captured and utilized, where energy is not only renewable but maximized through efficiency and sustainability. This is the promise that nanotechnology holds—a potential revolution in how we approach climate change.

The foundation of nanotechnology lies in the manipulation of matter on an atomic and molecular scale, typically less than 100 nanometers in size.

At this scale, materials exhibit unique properties that can be harnessed to develop new solutions to age-old environmental problems. The applications of nanotechnology are broad and varied, encompassing areas such as energy production, pollution reduction, and resource management.

Harnessing Energy with Nano-Materials

One significant area where nanotechnology stands to make an impact is in the realm of renewable energy. Solar power, often lauded as one of the most viable pathways to sustainable energy, benefits tremendously from advancements in nanomaterials.

New generations of solar cells are being developed using semiconductor nanocrystals and thin-film nanostructures, which offer higher efficiency levels compared to traditional photovoltaic cells. These advancements allow solar panels to capture and convert sunlight into electricity more effectively, even in low-light conditions or on cloudy days, thus maximizing their potential in diverse environments.

Moreover, nanoscale engineering extends to the realm of energy storage.

The creation of high-capacity batteries made possible by nanotechnology heralds a future where energy harvested from solar, wind, and other renewable sources can be stored efficiently and discharged on demand. Nanostructured electrodes can improve battery life cycles and charging times, addressing one of the critical bottlenecks in the broader adoption of renewables.

Air and Water Purification

Nanotechnology also offers groundbreaking solutions in pollution management. Air pollution, one of the most pressing environmental challenges, can be significantly mitigated using nanocatalysts and filters.

Nanoparticles such as titanium dioxide are employed to degrade pollutants in the atmosphere through photocatalytic reactions. These materials can effectively break down harmful substances like nitrogen oxides and volatile organic compounds, converting them into less harmful byproducts. In a practical sense, such technology could be integrated into urban infrastructures, such as building materials and road surfaces, to passively clean the air, particularly in densely populated cities.

Water, another critical resource strained by climate change, can be purified using nanotechnology.

Advanced filtration systems employ carbon nanotubes to remove impurities, heavy metals, and pathogens from water supplies. These systems are more effective and energy-efficient than conventional water treatment methods, and their scalability means they could provide clean water to remote and underserved regions worldwide.

Carbon Capture and Utilization

One of the boldest possibilities nanotechnology offers is in carbon capture and utilization (CCU).

While reducing carbon emissions remains crucial, the ability to capture and repurpose carbon dioxide from the atmosphere represents a pivotal advancement in climate change mitigation. Unlike traditional carbon capture methods, which often focus on sequestration, nanotechnology presents opportunities to convert captured carbon into valuable materials.

Nanostructured materials can be engineered to improve the absorption and conversion of CO_2 into hydrocarbons and other chemicals, which can then be used as industrial building blocks.

This not only aids in reducing atmospheric carbon levels but also turns a waste product into a valuable resource. Similarly, research into nano-enhanced photosynthesis, often referred to as "artificial leaves," mimics the way plants convert CO_2 into oxygen and glucose, providing yet another innovative pathway for carbon mitigation.

Minimizing Environmental Footprints

Moreover, the environmental impact of industrial and consumer products stands to be significantly reduced through the application of nanotechnology.

Developing environmentally friendly manufacturing processes and materials could minimize waste and minimize the carbon footprint of producing goods. Imagine self-cleaning fabrics, lighter and stronger building materials, or even packaging that biodegrades harmlessly—it all becomes possible with nanotechnology.

Enabling a circular economy, where waste becomes the fuel for further production, relies on the advancements made possible by nanoscale innovation.

Products designed at the nanoscale can often be engineered for a longer life span, higher durability, and enhanced recyclability, reducing the overall environmental burden and resource extraction required for new materials.

Challenges and Ethical Considerations

Despite its transformative potential, the role of nanotechnology in climate change mitigation is not without its challenges. The manipulation of materials at the nanoscale poses ethical and environmental questions.

Concerns over the toxicity and environmental impact of nanomaterials require careful oversight and regulation. Ensuring that these technologies are used responsibly and equitably is paramount as they become more integrated into our climate strategies.

Thus, the development of global standards and regulations governing the use and disposal of nanomaterials is crucial.

Collaborative efforts between scientists, policymakers, and stakeholders are necessary to navigate the ethical considerations associated with nanotechnology, guaranteeing that its applications are safe and beneficial to all.

A Call to Action

The role of nanotechnology in climate change mitigation is, without a doubt, a linchpin in our efforts to secure a sustainable future.

By embracing nanotechnology, we are not just finding new ways to mitigate the effects of climate change—we are unlocking opportunities for innovation that extend beyond energy and environmental management. The pursuit of such groundbreaking technologies inspires a broader reevaluation of how we interact with our planet and resources.

In conclusion, the integration of nanotechnology into our climate change strategies epitomizes a forward-thinking approach to environmental stewardship. It is not just about fixing problems but reimagining solutions, shifting paradigms, and creating technologies that lead to a more resilient and sustainable world.

Each nano-sized step is a monumental leap toward an enlightened era of green solutions, where the convergence of science, technology, and visionary thinking propels us toward a healthier planet. This is the role of nanotech in climate change mitigation—a testament to human innovation and an unwavering commitment to the future of our Earth.

Chapter 13: Health and Safety: Addressing Risks in Nanotechnological Applications

In the burgeoning field of nanotechnology, the promise of revolutionary eco-solutions looms large. From reducing pollution to creating sustainable materials, nanotech is poised to redefine our environmental landscape.

However, as with any powerful tool, the benefits of nanotechnology must be balanced with a thorough understanding of its potential risks. Specifically, health and safety concerns must be addressed proactively to mitigate potential dangers in various applications of this technology. This chapter delves into the critical facets of health and safety related to nanotechnology, illuminating the path toward safe and responsible innovation.

Imagine a world where invisible particles are implemented to detoxify polluted waterways, or where nanomaterials enhance the durability of infrastructure while simultaneously reducing their carbon footprint. As seductive as these prospects are, the unique properties that make nanoparticles so effective also render them unpredictable.

Their minuscule size — often between 1 and 100 nanometers — confers a higher reactivity, potential for permeability across biological barriers, and an increased surface area to volume ratio. These characteristics necessitate rigorous scrutiny, especially when they intersect with human and environmental health.

The first step toward conscientiously harnessing nanotech's capabilities is assessing its life cycle impacts. From production to disposal, each phase of a nanomaterial's life cycle poses distinct challenges and potential hazards. During the manufacturing phase, workers are at risk of exposure to airborne nanoparticles, which can penetrate respiratory systems more deeply than larger particles.

Inside the body, they may trigger inflammatory responses or distribute to organs such as the brain or liver. Thus, occupational safety protocols are paramount. Implementing proper ventilation systems, personal protective equipment (PPE), and comprehensive training programs can greatly mitigate these risks.

Next, consider the consumer phase. Nanotechnology-enhanced products range from cosmetics and clothing to electronics and food packaging. The latter, for example, might involve the use of nanocapsules that slowly release preservatives, extending the shelf life of perishable goods.

Yet, questions about the long-term impact of consuming such materials persist. Studies have shown that some nanoparticles can cross gastrointestinal barriers, potentially leading to cellular damage. Consequently, regulatory frameworks focused on rigorous testing and labeling requirements are essential.

Such measures ensure that consumers are not only informed but also safeguarded against any unforeseen health implications.

The environmental dimension of nanotechnology's health and safety is equally crucial. As nanoparticles are introduced into ecosystems, their interaction with biological organisms must be monitored. Studies have hinted that certain engineered nanoparticles can disrupt aquatic life by altering feeding, reproduction, and growth patterns.

The implications extend beyond individual organisms to entire food webs and ecosystems. Therefore, eco-risk assessments that model potential exposure scenarios are indispensable tools. These models help predict how nanoparticles might distribute in water, soil, or air, and guide strategies to limit ecological disturbances.

Believe it or not, Collaboration between scientists, industry stakeholders, policymakers, and the public is vital to fostering a balanced approach to nanotechnology's development.

Transparency in research findings and regulatory decisions builds public trust and facilitates knowledge exchange. International cooperation fosters standardized guidelines and safety protocols, which are crucial in addressing the global nature of nanotech applications.

Moreover, the paradigm of "safe by design" is gaining traction within the industry.

This proactive approach entails incorporating safety measures at the earliest stages of nanomaterial development rather than as afterthoughts. By selecting inherently less hazardous materials and engineering structures that minimize exposure or environmental impact, researchers can preemptively reduce risks.

Education also plays a key role in the responsible advancement of nanotechnology.

As the field evolves, training programs for scientists, engineers, and regulatory professionals must keep pace. Interdisciplinary curricula that blend technical expertise with ethical, environmental, and health considerations prepare emerging professionals to navigate the complexities of nanotech applications safely.

Yet, the quest for safety in nanotechnology is not without its inherent challenges.

The nanoscale domain operates at the intersection of quantum chemistry and classical physics, which means that traditional toxicological models may fall short in adequately predicting risks. Advanced research methodologies, such as high-throughput screening, in silico modeling, and novel in vivo studies, are needed to fill these knowledge gaps and refine risk assessment strategies.

Ultimately, the ethical imperative guiding nanotechnological innovation is not merely to avoid harm but to engender positive societal impacts.

Well-regulated nanotech offers transformative potential for addressing pressing global issues, from climate change to resource scarcity. By addressing health and safety risks with vigilance and innovation, stakeholders can ensure that nanotechnology upholds its promise as a pillar of sustainable development.

To this end, the chapter accentuates the need for a concerted effort to balance innovation with precaution.

The narrative of nanotechnology is not solely about pushing boundaries but also about respecting the limits imposed by health and environmental constraints. Through a meticulous approach to understanding and addressing these risks, the vision of a nano green future can be realized, where technology seamlessly intertwines with ecological harmony, enriching both the world and those who inhabit it.

Chapter 14: The Ethical and Social Implications of Nano Green Technologies

The dawn of the 21st century has heralded an era where technology intertwines with every aspect of our lives. At the forefront of this epochal shift stands nanotechnology, promising breakthroughs that transcend traditional boundaries.

When fused with green initiatives, this nexus gives rise to 'Nano Green' technologies – potent solutions poised to address environmental challenges. Yet, like any powerful tool, their deployment demands careful consideration of ethical and social implications.

Imagine a world where nano-structured materials in solar panels double their efficiency, bringing sustainable energy to the remotest corners of the globe.

Consider nanosensors embedded in agricultural fields that optimize water use and reduce pesticide reliance. While the potential benefits are captivating, they provoke essential ethical discussions about access, equity, and long-term impact.

Democratization vs. Disparity

The promise of Nano Green technology lies in its capacity to democratize access to essential resources. Solar cells with graphene layers could provide clean energy at a fraction of the current cost.

However, the path from laboratory innovation to widespread adoption is fraught with challenges. Who holds the keys to these technologies? Do they remain in the hands of wealthy nations and large corporations, exacerbating existing global inequalities? The ethical mandate is clear: proactive measures are crucial to ensuring these technologies do not become luxurious commodities but serve the common good.

It is imperative that international guidelines and partnerships foster a global distribution of Nano Green technologies.

Public and private sectors must collaborate to subsidize efforts in underprivileged regions, ensuring these initiatives cross geographical and socio-economic barriers. Otherwise, we risk engineering a future where the gap between the technological haves and have-nots grows wider, not narrower.

The Precautionary Principle

Nanotechnology holds immense promise, but its novel nature means that long-term effects remain uncharted territory.

The minuscule scale at which these technologies operate poses potential risks to human health and the environment. For instance, nanoparticles designed to clean water systems may inadvertently alter aquatic ecosystems or bioaccumulate in ways scientists do not yet fully understand.

This brewing uncertainty necessitates a rigorous application of the precautionary principle – the wisdom of forethought whereby the absence of complete scientific certainty is no reason to postpone actions that preemptively mitigate potential harms.

Regulatory frameworks must prioritize stringent testing and long-term impact studies, ensuring that Nano Green technologies fulfill their promise without endangering ecosystems or communities.

Transparency and Public Engagement

The introduction of Nano Green technologies into public life mandates an open dialogue among scientists, policymakers, and citizens.

Fostering a culture of transparency is crucial. Clear labeling of products containing nanomaterials, comprehensive data on their safety, and informed consent should be prerequisites for their market release.

Public engagement empowers communities to participate in decision-making processes, shaping directions that align with societal values.

Workshops, public consultations, and educational campaigns can demystify the science behind nanotechnology, transforming public perception from skepticism to informed curiosity.

Addressing Unforeseen Consequences

History teaches us that every technological leap bears unintended consequences.

The digital revolution linked global aspirations but also birthed privacy concerns and digital dependencies. Nano Green technology, while designing novel paradigms for sustainability, must remain vigilant of its shadow.

Consider the socio-economic impact if large sectors dependent on non-renewable resources face obsolescence due to efficient Nano Green alternatives.

The transition toward sustainable technology must address the livelihoods of those displaced, offering retraining programs and new economic opportunities to ensure a just transition.

As strange as it sounds, Moreover, environmental interventions at the molecular level might disrupt already fragile ecosystems.

Ecosystem resilience has limits, and a cascade of lost bio-resources could ensue from well-intentioned but ill-informed nano-enhancements. Ethical stewardship demands multidisciplinary collaboration, integrating the insights of ecologists, ethicists, and technologists to navigate these complex landscapes.

Conclusion: A Shared Future with Nano Green

The journey of Nano Green technologies is one of converging possibilities – bridging the chasm between invention and ethical stewardship, between progress and preservation.

As we stand at this crossroads, the choices we make will shape both the environmental tapestry of the Earth and the social fabric of humanity.

In an interconnected world, the implications of these technologies transcend individual domains, weaving a collective narrative etched in responsibility, foresight, and equity.

Ethicists and technologists alike must share the mantle of accountability, driving forward an inclusive dialogue that harmonizes innovation with human values.

It is not merely the advent of Nano Green technologies but our conscientious engagement with them that will define our legacy.

How we harness these microscopic wonders to serve the macrocosm of human and environmental welfare will determine whether we craft a future where technology uplifts society or, ominously, one where it loses sight of the very world it aims to save.

Chapter 15: Case Studies: Successful Nano Green Implementations Worldwide

In the ever-evolving landscape of technological advancement, nanotechnology stands out as a beacon of transformative promise, particularly in the realm of environmental sustainability.

Nano Green applications are not mere theoretical concepts, but rather, they are actively shaping a sustainable future across the globe. This chapter delves into several compelling case studies that illustrate the innovative application of nanotechnology in enhancing eco-solutions.

India: Nano-Filters for Clean Water

India, with its sprawling population and diverse geographical challenges, faces a significant hurdle in providing clean drinking water.

Traditional systems often falter due to cost and infrastructure complexities, but here, nanotechnology provides a beacon of hope. A notable project, spearheaded by a collaboration between Indian Institutes of Technology and local startups, has pioneered the development of nano-filters.

These filters, infused with silver nanoparticles, possess remarkable antibacterial properties.

Unlike conventional water purification methods, these nano-filters are cost-effective, require minimal maintenance, and are capable of eliminating up to 99% of contaminants including bacteria, viruses, and heavy metals. In a pilot initiative carried out in rural regions of Uttar Pradesh and Maharashtra, communities reported a drastic drop in waterborne diseases, underscoring the profound impact of nano-filters on public health.

Germany: Nano Coatings for Energy Efficiency

Germany, a leader in renewable energy adoption, has turned to nanotechnology to enhance energy efficiency in infrastructure.

A notable initiative involves the application of nano coatings on building exteriors. These coatings, made from engineered nanoparticles, provide superior thermal insulation and UV protection.

In a pioneering project in Munich, several residential complexes were equipped with nano-ceramic coatings. Observations over an 18-month period demonstrated a reduction in energy consumption for heating and cooling by 30%.

Moreover, the reflective properties of these coatings helped in mitigating urban heat island effects, a significant contributor to energy waste in metropolitan areas. This step not only contributes to individual cost savings but also aligns with Germany's ambitious carbon reduction goals.

Kenya: Nano-Pesticides and Agriculture

In real life, Kenya's agricultural sector is pivotal to its economy, yet it is constantly under threat from pests and diseases which traditional chemical pesticides cannot effectively manage.

Enter nano-pesticides: a groundbreaking solution that integrates nanotechnology with agriculture.

Local agricultural research institutes have developed pesticides that utilize silica nanoparticles as carriers for active ingredients. These nano-pesticides have improved pest penetration and delivery efficiency, allowing for reduced chemical usage without compromising effectiveness.

Trials conducted in the Rift Valley showed a marked increase in crop yields by up to 25%, alongside a significant reduction in pesticide runoff—a pressing concern for environmentalists and farmers alike. Not only does this lead to lush, healthy crops, but it also preserves the delicate ecosystem.

United States: Nanotech in Solar Panels

To be honest, In the quest to harness the abundant power of the sun, the United States has turned to nanotechnology to push the boundaries of solar efficiency.

Research initiatives at prestigious institutions such as MIT and Stanford have led to the development of solar panels utilizing perovskite nanoparticles.

These advanced panels are notable for their lightweight structure and flexibility, which allows for a broader range of applications, from rooftops to the surface of electric vehicles.

A case study involving a solar farm in Arizona demonstrated a 20% increase in energy conversion rates compared to traditional panels. This advancement signifies not just an incremental improvement but a leap towards making solar power more affordable and accessible, supporting the nation's transition to renewable energy.

Brazil: Nanotech in Oil Spill Remediation

It's worth noting that Brazil, with its expansive coastline and vast oil exploration activities, bears an inherent risk of oil spills.

Nanotechnology offers a promising solution to this environmental hazard. A collaborative endeavor between Brazilian universities and industry leaders has yielded nanomaterials capable of selective oil absorption.

These materials, crafted from magnetic iron oxide nanoparticles coated with oleophilic compounds, have the unique ability to separate oil from water.

A field trial in the aftermath of a spill off the coast of Rio de Janeiro showcased their efficacy, recovering 95% of the spilled oil in a fraction of the time required by traditional methods. This not only minimizes the environmental impact but also enhances the economic feasibility of cleanup operations.

China: Air Pollution Control with Nanotechnology

It's worth noting that In its ambitious pursuit to combat air pollution, China has turned to nanotechnology to purify its urban skies.

The city's governance in Beijing has implemented air filters embedded with titanium dioxide nanoparticles in public buildings and transit systems.

These nano-enhanced filters demonstrate photocatalytic properties, breaking down pollutants such as nitrogen oxides and volatile organic compounds into harmless byproducts.

Deployment across metro stations and public schools has led to a measurable improvement in air quality, with particulate matter levels dropping significantly. This success underscores nanotechnology's potential to safeguard public health amidst burgeoning urbanization.

Conclusion: The Global Impact of Nano Green Initiatives

Interestingly enough, These case studies illuminate the tangible benefits and diverse applications of nanotechnology in addressing environmental challenges worldwide.

From Asia to Europe, Africa to the Americas, Nano Green solutions are manifesting varied and profound impacts, championing a future where technology and nature coexist sustainably.

As we reflect on these successes, it becomes evident that nanotechnology is more than an innovative tool; it is a paradigm shift in our approach to environmental stewardship.

This worldwide embrace of Nano Green initiatives is fostering a global community dedicated to forging a cleaner, more sustainable world, thus paving the way for the generations to come.

In the spirit of progress and collaboration, let these examples inspire further exploration and adoption of nanotechnology in every corner of the globe.

After all, the quest for sustainability is not bound by borders, but united by a shared vision for a healthier planet.

Chapter 16: Challenges and Obstacles: Navigating the Nano Green Frontier

As we propel ourselves further into the 21st century, the fusion of nanotechnology and environmental sustainability has emerged as a groundbreaking arena with transformative potential.

Enthusiasts hail this alliance as the dawn of a new age, one where painstaking hurdles in ecological preservation find solutions in the infinitesimally small. Yet, this dazzling vision is laced with intricate challenges and formidable obstacles, demanding our innovation and perseverance.

The most pertinent challenge lies in understanding the complex nature of nanomaterials. At the nano scale, materials exhibit properties that fundamentally differ from their macroscale counterparts.

This unique behavior positions nanotechnology as a powerful tool for innovation, yet it simultaneously creates a landscape of uncertainty. Scientists and engineers face the daunting task of predicting and controlling the behaviors of novel nanomaterials, which, at times, defy conventional scientific paradigms.

I've seen this happen before: For instance, when considering the use of nanoparticles in water purification systems, researchers grapple with the complexities of ensuring particles efficiently bind with and neutralize contaminants.

While advancements are being made, the precise manipulation of these particles remains a formidable challenge. Furthermore, unforeseen interactions between nanomaterials and ecosystems or human biology could pose new risks, raising the need for comprehensive studies and stringent regulatory frameworks that can adapt to the pace of technological advancements.

Economic challenges are inextricable from the conversation. The initial costs associated with developing and implementing nano-enabled green technologies remain significant.

Despite the long-term savings and environmental benefits that these technologies promise, the scale-up from laboratory to market demands substantial financial investment. Moreover, the global nanotechnology market is dominated by entities from economically developed regions, posing an equity challenge. For the nano green revolution to be truly transformative, it must be accessible to nations across every rung of the economic ladder.

In parallel, public perception occupies a critical position in the nano green landscape. The general populace, unfamiliar with the subtleties of nanoscale science, often perceives 'nano' with skepticism or confusion.

The tiny scale, combined with its often intimidating scientific jargon, can seem like an enigma, concealing risks akin to those of powerful, unseen forces.

There is a pressing need for clear communication from scientists and policymakers to bridge this knowledge gap, ensuring that publics are both informed and engaged with nanotechnology's potential contributions to sustainability.

Believe it or not, Ethical considerations also loom large on the nano green frontier. As was the case with genetically modified organisms (GMOs) in the latter part of the 20th century, nanotechnology faces intense scrutiny regarding its moral and ethical implications.

The ecological impacts of mass-producing materials that might not degrade in the expected timeframe or might accumulate in unforeseen ways challenge our commitment to a precautionary principle. Moving forward, developing a robust ethical framework for nano green technologies will be indispensable in allaying public fears and ensuring responsible innovation.

Regulatory frameworks are another pivotal hurdle. Current laws and guidelines governing nanotechnology are often outdated, formed in times before the rapid acceleration of nanoscale research.

Crafting legislation that protects the public and the environment without stifling innovation is a delicate balancing act. Equipping regulatory bodies with the tools and expertise necessary to evaluate nanoscale products and interventions will ensure that we adequately address the complex interplay of risks and rewards posed by this nascent field.

Moreover, inter-disciplinary collaboration poses its own unique set of challenges. Nanotechnology inherently spans multiple avenues, from chemistry and physics to engineering and ecology, requiring synergistic collaborations across traditional boundaries.

Aligning these disciplines towards a common eco-friendly goal demands a paradigm shift in institutional practice and funding allocations. Collaborative efforts can unlock innovative solutions unforeseen by siloed approaches, yet they require considerable coordination, shared visions, and mutual respect among disciplines.

In real life, In conclusion, navigating the nano green frontier is no small feat. Challenges abound, but with them come opportunities to redefine our approach to sustainability.

Addressing these obstacles with creativity and foresight will not only bolster technological innovation but also ensure that our environmental stewardship marches hand in hand with scientific progress. As stakeholders—from scientists to policymakers, from businesses to educators—chart the course forward, the commitment to inclusivity, transparency, and ethical integrity will lay the groundwork for nanotechnology's full potential to realize a resilient, sustainable future. The true triumph in the nano green revolution will come not merely from the advancements themselves, but from the conscious, deliberate actions taken to integrate them into the fabric of our world responsibly and equitably.

Chapter 17: Policy Frameworks and Regulations: Governing Nano Green Solutions

In the dynamic intersection of nanotechnology and ecological sustainability, the role of governance cannot be overstated. The dawn of 'Nano Green' solutions presents a promising horizon in our battle against environmental degradation, yet it simultaneously poses unique challenges that necessitate innovative policy frameworks and regulatory architectures.

In this chapter, we shall explore the labyrinthine world of policies and regulations that the nano green revolution navigates. Our journey will unravel the intricacies of governmental oversight essential for safeguarding public interests while encouraging technological progress.

To appreciate the necessity for specific policies in nano green solutions, one must first understand the transformative potential of nanotechnology in environmental contexts. Unlike conventional technologies, nanotech operates at the molecular level, promising miniscule, yet monumental impacts.

These include the remediation of pollutants, the enhancement of energy efficiency, and the sustainable management of resources. But it is this very potential for profound change that compels policymakers to tread carefully; ensuring that nano solutions flourish without becoming dystopian catalysts of new environmental or health crises.

It's worth noting that Globally, the regulation of nanotechnology remains in nascent stages, often grafted onto existing policy structures without fully addressing the nuanced needs of nano-specific applications.

This lack of dedicated frameworks can engender regulatory gaps where oversight is critical. For example, many countries still apply chemical safety standards, developed with bulk materials in mind, to nano-enabled products— a misstep considering that nanoparticles often behave differently than their larger counterparts.

The first step towards a coherent policy framework is recognizing the dual nature of nanotechnology as both a tool for innovation and a potential risk. Governments play a pivotal role here; tasked with not only propelling research and development (R&D) through funding but also establishing safety and ethical guidelines that prevent misuse.

National strategies for nanotechnology are emerging, aiming to balance these aspects. For instance, the European Commission has laid out comprehensive frameworks that guide member states in integrating nanotech considerations into environmental policies, such as the REACH regulation, which obliges the chemical industry to provide data on both new and existing substances, including those at the nanoscale.

Moreover, the need for international cooperation and standardization cannot be overlooked. The cross-border nature of environmental issues, combined with the globalized economy, means that isolated policies may fall short.

Institutions like the International Organization for Standardization (ISO) and the Organisation for Economic Co-operation and Development (OECD) are crucial. They set standards that harmonize practices and facilitate the safe exchange and application of nano green technologies worldwide. By establishing uniform definitions and measurements, these bodies help mitigate discrepancies in regulatory practices that might otherwise hinder international collaboration or trade.

Interestingly enough, However, developing effective regulations requires more than just scientific data and international agreements; it necessitates collaboration with stakeholders from various sectors. Public participation, industry input, and academic insight should form the trinity of an inclusive policy-making process.

Engaging these groups garners a diversity of perspectives, ensuring that policies are robust, adaptable, and equitable. For instance, participatory approaches have proven effective in fostering trust and acceptance of new technologies, which is critical in avoiding public skepticism and enhancing societal benefits.

In addition to collaborative policymaking, adaptability is a cornerstone of effective governance in the rapid-tech field of nanotechnology. Regulatory systems need to be forward-thinking and reflexive, able to evolve alongside technological advancements.

Adaptive policies, often encapsulated in the doctrine of "regulatory sandboxes," allow for real-world experiments under careful oversight, enabling policymakers to observe potential impacts in situ. This is vital for nano green technologies, where real-world dynamics can differ significantly from laboratory predictions.

Despite these progressive approaches, challenges remain. A significant issue is the limited availability of comprehensive risk assessments.

Nanotoxicology is still developing as a scientific discipline, and data deficiencies impede the understanding of potential long-term effects on both health and ecosystems. Policymakers must therefore support further research in this area, ensuring that regulations are informed by robust evidence.

It's worth noting that Moreover, ethical considerations play a crucial role. The prospect of manipulating atoms and molecules carries with it moral responsibilities.

Policymakers must address such concerns, ensuring that nano green innovations align with broader societal values. Questions surrounding privacy, equity, and access to technology must be part of the regulatory discourse. Ethical frameworks must elevate human and environmental well-being as central tenets, guiding the ethical deployment of nano solutions.

Ultimately, the symbiotic relationship between policy frameworks and technological innovations in the nano green revolution is undeniable.

As we navigate this new frontier, the challenge lies in crafting policies that are neither overly restrictive—hindering innovation—nor too lenient, risking unforeseen consequences. It requires a delicate balance, integrating advanced scientific understanding with principled governance, ensuring that as we pivot towards sustainability, we do so with the foresight and precaution that a responsible transition demands.

Thus, the narrative of policy frameworks and regulations in governing nano green solutions is one of intricate interplay. It is a tale of balance—between innovation and caution, between the local and the global, between the known and the emerging unknown.

As we stand at the precipice of unprecedented environmental challenges, our ability to harness nanotechnology responsibly will be a testament to our collective wisdom and ingenuity. As such, the journey to drafting and implementing effective policy frameworks is as critical as the technological innovations they aim to govern, promising a future where environmental sustainability and cutting-edge science walk hand in hand.

Chapter 18: The Economics of Nano Green: Assessing Costs and Benefits

In the dawn of any technological revolution, the conversation inevitably swings towards the economic impact. With the advent of Nano Green technologies, the synergy of sustainability and nanoscale innovations, this discussion becomes particularly salient.

Nano Green promises unprecedented advancements in all aspects of environmental management and industry, from energy efficiency to pollution reduction. However, assessing its economic viability demands a nuanced exploration of costs and benefits.

The Cost Structures of Nano Green Technologies

As strange as it sounds, While the allure of Nano Green innovations lies in their potential to create sustainable futures, understanding their cost structures is essential. At the heart of these structures are research and development (R&D) costs. Developing nanotechnologies demands investment in specialized equipment and expertise.

Laboratories equipped with state-of-the-art facilities to analyze materials at the atomic level don't come cheap. Coupling this with the need for sustained innovation to maintain competitive advantage, initial investments can be significant. Yet, these costs are often mitigated over time by the economies of scale achieved as technologies mature and market adoption increases.

On the production side, manufacturing nano green solutions presents unique challenges. Producing nano materials typically requires specialized processes different from traditional manufacturing protocols.

Nanoparticle synthesis, for instance, often involves precise temperature control and sophisticated dispersion techniques. This necessitates an initial outlay for refining and optimizing production lines. However, the silver lining is that once production processes are streamlined, material efficiency often improves dramatically, leading to lower per-unit costs compared to traditional materials.

Yet, nano green technologies benefit from a dual financial character. They hold the potential to directly tap into savings through material reduction—achieving more with less.

Take, for example, nano-enhanced solar panels that use minimal coating materials while maximizing light absorption efficiency. Similarly, nano-coatings on packaging materials can significantly reduce the need for single-use plastics by improving durability and shelf-life.

Many people don't realize this, but ## The Benefits of Implementing Nano Green Solutions

Economically, the allure of Nano Green goes beyond simple cost savings—it heralds a transformation in value propositions. The environmental benefits delivered by these technologies translate directly to economic advantage.

For instance, companies that embrace nano green solutions can achieve major gains in operational efficiency. Improved energy efficiency from nano-enhanced materials reduces operational costs, a critical feature in an era when energy pricing is volatile and carbon footprints are scrutinized. Additionally, the lifecycle of products is extended through improved durability and resistance to environmental degradation, leading to further long-term savings.

Investor interest in nano green solutions is another significant economic benefit. As society increasingly prioritizes sustainability, green technologies attract substantial investment capital.

Venture capitalists and forward-thinking enterprises see lucrative opportunity in funding the development of technologies that simultaneously fulfill environmental and economic criteria. In this context, Nano Green emerges as a compelling narrative of investment with tangible returns.

Believe it or not, Moreover, the adoption of nano green technologies plays into broader economic trends, such as government policies and consumer preferences.

Policy incentives, including subsidies and tax breaks for companies that implement sustainable business practices, amplify the attractiveness of these investments. Additionally, consumer demand for green products, now a defining feature of markets worldwide, ensures a ready market for nano-enhanced goods.

Balancing Initial Costs with Long-term Gains

A thorough economic assessment involves balancing the immediate costs with long-term gains.

This dynamic is particularly pronounced in nano green technologies where initial expenses must be evaluated against savings achieved over time. Life-cycle analysis becomes crucial here, assessing both direct and indirect economic impacts.

At first glance, Direct impacts include tangible savings like lower energy bills and reduced material wastage. Indirect impacts, however, carry equal if not greater significance.

These encompass brand reputation, customer loyalty, and compliance with regulatory frameworks. In industries where environmental consciousness becomes a determinant of market share, early adoption of nano green solutions can serve as a distinguishing factor. Corporations embracing innovations not only adhere to emerging regulatory landscapes but also position themselves strategically within the marketplace, reaping benefits of enhanced consumer trust and reduced future compliance costs.

A pivotal aspect of this balance is understanding technology readiness levels (TRLs).

Many nascent nano technologies remain at lower TRL stages, thus offering more speculative long-term gains but potentially greater returns on investment. As technologies ascend on the TRL scale from laboratory validation to real-world application, initial financial risks decrease concurrently with uncertainty, yielding increasingly stable economic forecasts.

The Role of Collaborative Economics

Let's not forget, Economic considerations of Nano Green cannot exist in isolation; they thrive on collaboration. Strategic partnerships address the twin challenges of cost and innovation.

Collaborations between academia and industry, for example, can lower R&D costs and accelerate the transition from theoretical exploration to commercial exploitation. Academic labs provide a fertile ground for experimentation, allowing industry partners to test innovative solutions without bearing the full brunt of R&D expenses.

Moreover, consortiums of corporations pooling their efforts can capitalize on shared learning and mitigate market risks.

Joint ventures enable effective resource allocation and foster environments where breakthroughs aren't just possible, but probable. This collaborative economics not only presents shared cost-savings but enriches competitive landscapes by setting industry benchmarks for future nano green advancements.

A Global Economic Perspective

Let's not forget, Addressing the economic dimensions of Nano Green extends globally, recognizing the disparity between developed and developing regions. Industrial nations with established nanotechnology infrastructures may achieve swifter cost-benefit equilibria, while developing nations confront additional barriers.

However, the universal imperative of sustainable development offers a unique platform for international cooperation. By sharing technology, knowledge, and economic strategies, we forge a path where Nano Green becomes an accessible asset rather than an exclusive privilege.

In conclusion, the economics of Nano Green uncovers a complex yet promising landscape. It challenges preconceptions about costs, urging stakeholders to envisage a future where economic growth and environmental stewardship progress symbiotically.

While hurdles remain, the transformative potential of Nano Green solutions introduces an era of sustainable prosperity. As we embrace these innovations, we not only contend with economic realities but redefine them, crafting new paradigms for a world entrenched in collaborative growth and ecological sustainability.

Chapter 19: The Future of Nano Green: Innovating for a Sustainable Planet

In the vast symphony of technological advancement, where each innovation holds the potential to either uplift or burden our planet, nanotechnology stands out as a herald of possibility—a keystone in the archway to a sustainable future.

"Nano Green," a term that once resided solely within the annals of science fiction, now strides confidently into the realm of reality, offering solutions that epitomize ecological harmony.

As we peer into the coming years, the horizon of Nano Green promises not only to enhance our existing capabilities but to fundamentally reshape how we interact with our environment.

The very essence of nanotechnology lies in its ability to manipulate matter at an atomic scale, providing an unparalleled opportunity to redesign traditional approaches to sustainability.

Harnessing the Power of Nature

Imagine a world where our interactions with nature are not plagued by extraction or consumption but are defined by collaboration. Nanotechnology has the potential to mimic natural processes, leveraging the intrinsic efficiency honed by millions of years of evolution.

Consider photosynthesis—the quintessential natural process that efficiently converts solar energy into chemical energy. By developing nanoscale catalysts that replicate this process, scientists aim to revolutionize solar energy harvesting, creating photovoltaic cells that are not only vastly more efficient but also more adaptable and affordable.

Purification and Remediation

To be honest, Water, the lifeblood of our planet, stands at the center of ecological concerns. Nano Green proposes radical advancements in water purification systems that could render access to clean water a ubiquitous reality.

By employing nanomaterials with extraordinary absorptive capacities, these systems can remove contaminants with surgical precision. Moreover, nanotechnology enables the design of membranes that are both selective and sustainable, facilitating the desalination process without the intensive energy consumption that current methods require.

In tandem, nanotechnology offers innovative solutions for environmental remediation. Consider the daunting challenge of oil spills—conundrums that have historically left a trail of devastation in their wake.

Nano Green envisions nanoparticles designed to attract specific pollutants, effectively acting as microscopic magnets that bind to contaminants, rendering them harmless. This method not only promises efficiency but also echoes a greater respect for the biosphere.

Revolutionizing Agriculture

As strange as it sounds, One of the most critical fronts for sustainability lies in agriculture—an arena primed for the transformative touch of nanotechnology. The Nano Green approach to agriculture could redefine conventional farming, with nanoparticles offering precision delivery systems for nutrients and pesticides.

Envision nanosensors embedded within soil, providing real-time data on moisture and nutrient levels. These sensors can guide farmers to make informed decisions, minimizing waste and maximizing yields in a manner that remains attuned to the delicate balance of ecosystems.

Furthermore, novel nanopesticides and fertilizers could significantly reduce the chemical burden on our environment.

With the ability to target pests and plants with precision no larger molecule could achieve, these nano-formulations promise to uphold crop integrity while curtailing collateral damage to surrounding flora and fauna.

Energy Solutions at the Atomic Level

As strange as it sounds, The energy sector is poised for a Nano Green revolution. Nanotechnology's potential to enhance energy storage and conversion capabilities heralds a future where our dependence on fossil fuels could dramatically diminish.

Batteries constructed with nanoscale components boast increased capacity and faster charging times, paving the way for electric vehicles that can traverse greater distances on a single charge. These advancements are not merely incremental; they are exponential, redefining the limits of what is possible.

Parallel to storage innovations, nanotechnology in fuel cells could bring hydrogen energy into sharper focus, offering a clean, efficient, and sustainable power source.

Nanocatalysts capable of operating under ambient conditions would drastically simplify hydrogen production and consumption, making it a viable everyday energy solution.

A Circular Economy Reinforced by Nano

At first glance, In the journey toward sustainability, the circular economy model—where waste is systematically repurposed—finds an invaluable ally in nanotechnology.

Nano Green innovations facilitate the breakdown and reconstruction of materials at a molecular level, allowing a seamless conversion of waste back into valuable resources. For instance, nanoscale catalysts could help in breaking down and recycling plastics more efficiently, mitigating one of the most pressing environmental challenges of our era.

Human Health and Environmental Safety

The promise of Nano Green is not without its responsibilities. As we advance into a future suffused with nanomaterials, ensuring their safety for both human health and the environment is paramount.

Rigorous testing and transparent regulation must accompany innovations, paving the path for responsible implementation. With careful stewardship, the potential risks can be mitigated, transforming nanotechnology from a promise into a pillar of sustainable progress.

Conclusion

The future, as envisioned by Nano Green, is radiant with potential—a shimmering landscape where human ingenuity not only coexists with the natural world but thrives within it.

Through the lenses of nanotechnology, our capacity to forge sustainable solutions becomes boundless, reimagining agriculture, energy, water, and waste. As Nano Green innovations transcend the titular boundaries of theory and orbit into practice, the trajectory toward a sustainable planet accelerates, propelled by the smallest of scales yet offering the grandest of impacts.

In this emergent era, we stand on the precipice of not merely a new chapter, but a new paradigm—one where the planet and its people prosper with synchronicity, guided by the invisible but indelible hand of nanoscale innovation.

The future of Nano Green is not a distant dream; it is an imminent revolution, beckoning us to step forward with foresight, courage, and an unyielding commitment to the sustainability of our shared home.

Chapter 20: Conclusion: Embracing Nanotechnology for a Greener Tomorrow

As we approach the conclusion of our exploration into the profound confluence of nanotechnology and environmental stewardship, we stand at the precipice of a future laden with both promise and responsibility.

The intricate dance between these two fields has unlocked avenues that were previously mere figments of scientific imagination, presenting us with transformative avenues to address the ecological challenges of our time.

Nanotechnology, with its capacity to manipulate matter at an atomic and molecular scale, has emerged as a cornerstone for innovative eco-solutions.

The very essence of this technology lies in its ability to create materials and devices that are incredibly small yet remarkably potent, allowing for unprecedented precision in a myriad of applications. From water purification to renewable energy and waste management, these nanoscale interventions are designed not only to reduce environmental footprints but to actively heal the wounds we've inflicted upon our planet.

As strange as it sounds, However, the adoption of nanotechnology in our quest for a greener tomorrow requires an earnest commitment to scalability and sustainability. The miracles of nanoscale science need not be confined to the laboratories or niche products; rather, they must become integral to global developmental and industrial frameworks.

This necessitates an ecosystem that encourages interdisciplinary collaboration among scientists, policymakers, and entrepreneurs. Only through partnerships that foster innovation and pragmatic deployment can the true potential of nanotechnology be realized.

Consider, for instance, the burgeoning innovations in solar energy. Nanoscale photovoltaic cells that boast enhanced efficiency and reduced production costs stand poised to transform the way we harness solar power.

Yet, the challenge lies not just in perfecting these technologies but in disseminating them widely, ensuring that they reach communities worldwide. By democratizing access to such clean technologies, and integrating them into mainstream energy systems, we could accelerate the transition to renewable energy, reducing reliance on fossil fuels and decreasing greenhouse gas emissions.

Similarly, the revolution in water purification through nanomaterials signifies a beacon of hope for millions lacking access to clean water. Innovations like graphene oxide filters capable of removing contaminants at an astonishing rate are more than scientific milestones; they are life-changing enhancements to health and wellbeing.

The impact of such technology hinges on cost-effective production and distribution strategies. By reducing barriers to access, this technology can fortify efforts to combat global water scarcity, fostering resilience in regions plagued by drought and pollution.

To be honest, As inventors and innovators push the boundaries, they must not lose sight of the ecological tapestry within which these technologies will operate. Biocompatibility and biodegradability must be non-negotiable attributes of nanotechnological products.

The development process should rigorously evaluate potential ecological impacts, ensuring that today's solutions do not become tomorrow's problems. Conducting comprehensive life cycle assessments and maintaining stringent regulatory frameworks are imperative to balance innovation with environmental sanctity.

Moreover, public perception and education will play pivotal roles in the widespread acceptance and success of nanotechnology. Awareness campaigns and educational programs can dismantle the wall of skepticism that often accompanies emerging technologies.

By demystifying nanotechnology, the public can engage more robustly in its discourse, contributing to a transparent flow of ideas that cultivate trust and accountability. It is through shared awareness that societies can adopt these innovations not with apprehension, but with informed enthusiasm.

The path to a greener tomorrow demands that we reimagine traditional industries through the lens of nanotechnology.

Agriculture, with its ever-growing demand for efficiency and sustainability, stands to be revolutionized through nanoscale agrochemicals and smart delivery systems that ensure nutrients and pesticides are applied precisely where needed, reducing waste and runoff.

Many people don’t realize this, but Meanwhile, in the realm of waste management and recycling, nanotechnology's ability to deconstruct materials to their elemental components presents a new frontier in material recovery and reuse.

Innovations in nanoscale sensors and sorters could transform what we perceive as waste into raw materials for new products, promoting a circular economy.

Ultimately, embracing nanotechnology as a catalyst for ecological change will require more than technological advancements; it will demand a paradigm shift in how we envision and create solutions.

This revolution is as much about shifting mindsets as it is about developing new materials.

In the quest for a sustainable future, nanotechnology presents a set of tools so powerful and versatile that the scope of their application is as expansive as our imagination.

Yet, with great power comes a profound moral imperative: to wield these tools in the service of humanity and the environment. The legacy of this era will be defined not only by the innovations we forge but by how we choose to use them—responsibly, equitably, and sustainably.

In real life, As we conclude this exploration, let us cast our gaze toward the horizon with optimism and resolve.

The intersection of nanotechnology and green innovation marks the dawn of a new epoch, one where the dreams of a healthier planet can finally take shape. Let us embrace this opportunity with courage, cultivating a future where technology and nature coexist harmoniously, and where our children inherit a world defined by its ingenuity and care for the Earth.

In doing so, we join in a collective journey—a journey not only toward progress and prosperity but toward wisdom and stewardship.

For in the tiny particles we now command, we hold the potential to sculpt the landscapes of tomorrow, a world where ingenuity knows no bounds, and sustainability is not merely an ideal but a lived reality.

About the Author

T. S. AVINI is a writer who believes in the power of sharing life's lessons. Having faced hardship and pain, he turned to writing as a form of self-expression and healing. Today, he pours his heart into every book, aiming to help others navigate their own struggles and find strength in vulnerability.